KaBaFa

Melanin Redefined

Authored by

T. Owens Moore, Ph.D.

Z

Zamani Press
P.O. Box 667
Redan, GA 30074
ZamaniPress.com

KR
PRINTING & PUBLISHING
KeithRyales.com

Front cover Illustration by Keith Ryales

First Edition, First Printing

ISBN#: 978-1-884897-06-1

Table of Contents

Dedication

To the contemporary melanin scientists who have transitioned to the Amenta World. Without their scholarly contributions to the topic, we would be less "endarkened."

Laila Afrika
Anthony Kweku Andoh
Malachi Andrews
John Chisel
Richard King
Pat Newton
Francis Cress Welsing
Amos Wilson
...and others now known in Amenta.

Acknowledgements

I first acknowledge the living legends who brought me to this place and space to think about the mystery of melanin. Drs. Na'im Akbar and Wade Nobles are two of four authors from the 50-year-old classic article, Voodoo or I.Q.: An Introduction to African Psychology (Clark (X), McGee, Nobles and Weems, 1975). Much respect is given to these two Jegnas for being a foundation in my academic life and for raising the topic of melanin as a critical factor to consider in human behavior.

Also, to my contemporary African-centered scholars like Jeff Menzise, Mark Bolden, Michele Lewis, Lennell Dade, Chike Akua, Niyana Rasayon, Olisa Tolokun, Marilyn Lovett, Shawn Utsey, Jewel Pookrum, Bruce Bynum, Hunter Adams, Ann Brown and too many others to mention, I appreciate the opportunity to bounce ideas off your consciousness.

To my university colleague, Medha Talpade, you are greatly appreciated for introducing our students to my work. I value you for valuing my work. And to get my work right, I thank Vikki Johnson for your editorial assistance. To my current and past students, you are the reason scholars like me are driven to provide knowledge. Thank you for creating the thirst.

To Nia Damali, the owner of MEDU Bookstore, thank you for the support over the years. Without you, the ATL would be void of an intellectually stimulating environment to learn and love books. Along with your bookstore, I am thankful for the Black Dot Cultural Center owned by Kazemde Ajamu and Wekesa Madzimoyo and the AYA Institute. You will always be appreciated and thank you for supporting my work. All three of you are valued members of the community.

Many thanks to my comrade and soul brother Lynn Hampton for allowing me to present to the Cleveland crew and to my soul sister Orisade Awodola and her supportive conferences in Seattle, WA. Also, I am thankful to the international community who supports my work through purchases on the website, and I am honored you share an interest to promote this message about melanin.

Finally, and not least, to all my immediate family and close community friends, you will always be the foundation for why I live by the principles of MAAT.

Preface

This is my sixth book on melanin, and I feel compelled to reevaluate how we perceive the topic in public discourse. I first wrote about it three decades ago because it was controversial. Times have not changed, and the pigment politics associated with the topic lead me to address some concerns so we can be more accurate on our conceptualization of what we mean by saying melanin to describe what "black" people have.

What I write here does not change what I have written in the past. There is no retraction in what I have previously expressed about melanin, but it is a moment in history that calls for a redefinition of terms. I first presented this topic on KABAFA: Melanin Revisited at the 2023 Association of Black Psychologist's International Convention in Detroit, MI, and I requested permission from the elders, the members attending my session, and the audience participants to change the name of melanin to KABAFA as a term to go forward. Therefore, this alternative view of expressing melanin was granted by a jury of my peers to use KABAFA as an explanation to describe melanin as the conduit between the material and spiritual connection in the human experience.

To say Melanin Revisited is a copied theme from Ivan Van Sertima's edited book called Egypt Revisited (1989). His foundational text gave us a new view on an old topic, and it helped to provide a foundation on how we can reinterpret and correct past history that was told from a view that was antithetical to African-centered thought.

This is my shortest treatise on the topic, but it is critical of how melanin has been used, and I focus on why a change is inevitable as we advance knowledge and intellectually grow.

In Chapter 1, the Science of Spirit, I express what led me to address this topic on melanin revisited and it was the death of close associates. Their physical bodies were gone, but their presence was still here.

In Chapters 2, 3 and 4, I set the foundation for the respective terms KA, BA and FA and why they were chosen. In Chapter 5, I reveal that I am not the originator of these thoughts on fire atoms. Our ancient African ancestors had terms that I believe described "melanin" before the Greeks gave us this word. We present information on the driving term called fire atoms that demonstrates a higher thinking on concepts related to the collective unity of how cells work together to give us higher states of consciousness.

In Chapter 6, Soul Studies, I emphasize the need to create new disciplines to explore these topics that have been known to be immeasurable.

In the final Chapter 7, I collectively summarize the sun as the progenitor of all life and how KABAFA emanates from the Flesh of RA (i.e., the sun). Self-knowledge is the key to life, and the epilogue provides an example of this importance to KNOW THY SELF.

Introduction

A critical element of life - that is what we are referring to when we think of KABAFA = melanin. Melano – comes from the Greek melas, which means black. Rather than using Greek words to establish African-centered terminology related to darkly pigmented people on the planet, we should utilize terms that define the heritage of the people being described. Ka-body, Ba-breath, Fa-fire atoms are where the term is generated. KABAFA becomes the essential element of life, and it can be represented in many colorful shades. Pigment is what we are referring to and this is commonly known as melanin in western vernacular. This pigment is a complex biopolymer that gives coloration as well as unique hueman experiences.

The variety of pigment or coloration in the feathers of birds, the color of insect bodies, fur of mammals, scales of invertebrates, and the color on mold and fungus found in nature is NOT what we are referring to as KABAFA.

KABAFA is the essence of how the darkness in the skin, eyes, ears, hair, internal organs, and brain of humans respond to the frequencies of nature. This physical connection to nature is the key element that makes the experience of being a "hueman."

If we reanalyze the convenient use of melanin-dominant and melanin-recessive to describe people, we can eliminate the use of melanin as a term to define differences in humans. Instead, KABAFA emphasizes that all people have potential to experience the frequencies of life, but the level of KABAFA can affect your overall awareness of what it means to be human and to be connected with higher realms of consciousness.

Let us be clear, black does not justify the use of our conceptualization of melanin because people with melanin may not be "black" in appearance. As we can see in history, the number of cultural classifications pertaining to melanin-dominant bodies indicates a social quandary (e.g., mulatto, mixed, octoroon, quadroon, preto, negro, etc.). Even the use of the social construct called race adds to the confusion because there is no black, white, yellow, red or any other color that should be classified as a race.

Since people express different shades of blackness from their external appearance, we cannot simply associate the complexion of a human body as the difference maker. We cannot say that light-skin is better or worse than dark-skin on a psychospiritual level. It is the overall genetic capacity to produce melanin (i.e., KABAFA) that provides a person with the ability to fully experience a psychospiritual consciousness that is connected to nature.

It is the machinery in the genes and what is held in the cell that is magnified when we are studying the science of the spirit.

The term KABAFA is a way to redefine our outlook on the all-encompassing topic of melanin. To reiterate, the role of melanin is confusing when you only associate it with pigmentation. Yes, it provides various colorful pigments, but the concept goes beyond color. We can be confused because the color story leads to colorism, racism and division. We need an alternative outlook on what is meant by blackness (Gates, 2024) when we describe people and melanin.

There is a need to truly address the African-centered view of studying people who are melanin- dominant. There are significant flaws in descriptors when the people who are defining the essence of something do not possess what is being described. Henceforth, we see a reason why the soul and spirituality cannot be clearly expressed from a western or Eurocentric viewpoint. The soul and spirit are metaphysical concepts that emanate from the properties of what we call melanin. The only obvious connection to these terms from a

Eurocentric analysis is to confine the concepts into what is known as religion. Religion is NOT spirituality, and the structure of religion can actually decimate a person's soul. Orienting yourself to a God-like image that does not look like you is not salvation, especially if you view this image as a melanin-recessive person.

We can be myopic in our view and think that an understanding of the soul and spirituality are only taught from one perspective. The major religions such as Christianity, Islam, Buddhism, Hinduism have had a major influence on world thought, but we must acknowledge that Africa is the mother of Western religions (Ben Jochannan, 1970). In conceptualizing indigenous African religions, one cannot begin to speak of man/woman without first speaking of this force which defines his continuity with all things within the world.

For example, the Dogon people refer to this force as nyama, which flows with the blood in man's veins and connects his actions and circumstances with the functioning of all things in general. The Bambara people call it dya, which serves to unify all things in nature. The Akan refer to it as a man's kra – his life force. The BaNtu-Kongo call it kalunga, which is fundamentally a process of perpetual and mutual sending and receiving of spirit (energy) in the form of waves and radiations. As we will explore throughout this text, this notion of such a universal force is most closely rendered by people of African descent throughout the diaspora as the soul.

Any non-Eurocentric outlook focuses on the elevation of the soul and spirit. Melanin-dominant people are those who would express an elevation of soul and spirit in their cosmology.

To emphasize this conceptual framework, it will require an investigation of ancient terms and the advancements in science to reconceptualize what we mean by melanin, blackness, darkness and/or pigmentation. I suggest that the new conceptualization be called KABAFA.

CHAPTER 1

The Science of Spirit

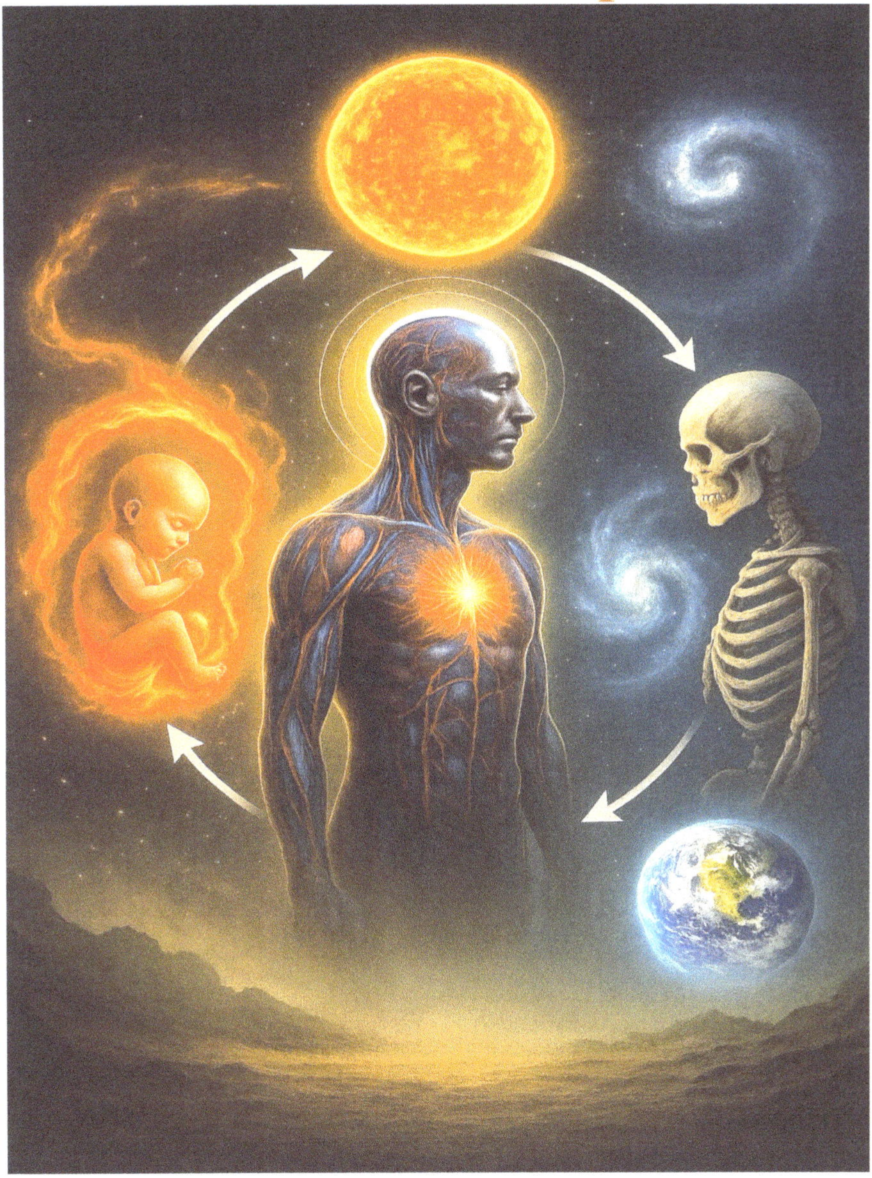

Fig. 1 The birth, life and death cycle manifests on the physical plane for us to experience a spiritual connection to the universe.

Death has led me here. Two of my best friends passed away at the tender age of 73 in 2022. They went onto the ancestral realm, and I shed tears, but I was immediately aware that they were still "here." In other words, I consciously wanted to avoid the obvious that they were gone,but their physical body was lifeless, and the experience was death. Their bodies changed and transformed into another realm, but their spirit was in the air for me to know they existed. These friends were not blood relatives and there was no genetic connection. However, reflect on your family lineage and observe the birth-life- death cycle. In other words, when a relative physically dies, almost within a year time period, someone is born within your family lineage.

What destroys this connection in families is the fractured social relations many people have where families do not know one another anymore. It is, therefore, difficult to keep up with who is in your family. There is so much disconnect that many of us do not even know where our family members are located. Under these circumstances of not knowing family lines, we will have difficulty keeping up with the science of spirit. What is reflected here as a problem to understand the science of spirit was a technique that was mastered by our ancestors.

For example, the cultural customs and rituals in ancient Kemet exemplify a civilization that understood the science of spirit. The power of metaphysical concepts related to spiral energy, mummification and multiple divisions of the soul are clear evidence of a sophisticated civilization.

The science of spirit must be understood to process the way KABAFA is manifested in the human experience. The body can be lifeless, breath can be gone, and the heat generated from the internal electricity can be extinguished. As we will describe in the next chapters, KABAFA is the combined analysis of the physical body, the delicate breath and the electrical energy as understood by ancient African scientists.

In death, this is nothing but a change in frequency for what the hue in the body receives. Yes, the heart stops beating, and the breath is gone, but the spirit is still present on another level of the electromagnetic spectrum. The ancient Africans lived for the afterlife and by living by the principles of MAAT, life would be spiritually connected between the present and future. Like matter cannot be created or destroyed in physics, spirit is eternal. This is why in war, the spirit of a people will conquer all flesh.

To avoid death, you must be healthy and well. What does wellness entail? To be well, you must have control over your mind, body and spirit. The mind is first, the body is second, and the spirit is third.

KABAFA is related to all three aspects of our life experience. First, the mind tells us what we should know, and it guides us to keep a healthy body intact. Without the mind, we would not know how to conceptualize body functions or to determine what spirit means. Cellular groups in our brain store information to allow our mind to work correctly. If those neuromelanin containing cells or any pigmented cells do not work, then a person's consciousness will be impacted, i.e., your memory could fail, and you might lose your mind.

Although second, the presence of KABAFA in the body is critical for life. KABAFA on both the inside as well as the outside of the human body serves to keep the body healthy. KABAFA can neutralize toxic substances and prevent cellular damage, so the body is enhanced when KABAFA is functioning at an optimal level.

Not that it is the least important, but it may be more difficult to conceptualize how the spirit is manifested.

If we think of spirit as all-encompassing and an integral part of nature, we can understand the connection with the KABAFA in our body, in living organisms, and in the Cosmos. Spirit is everywhere and it pervades the entire electromagnetic spectrum. Therefore, KABAFA is the conduit between the material and spiritual realms of existence.

Pre-Melanin Cosmology in the Grand African Narrative

The purpose for the accumulation of knowledge is to advance intellectual inquiries and to enhance human development. At the forefront of studying, contemplating and formulating new conceptualizations of spirit, we can look no further than exploring the legacy work of Dr. Wade Ifagbemi Sangodare Nobles. In his life-long academic search in seeking the Sakhu (2006), he provides a greater understanding of the importance of KABAFA that is presented in this book.

In the eternal fight for the liberation of the African Spirit, he has laid a foundation for our contemporary application of ancient wisdom. Just as the emphasis is placed on Kemet in this book, Nobles explores the many connections of the African mind extended over historical time. In his book called SKH: From Black Psychology to the Science of Being be states that "this offering will allow the reader to wade in the ancient Nile River of thought, shower with cool waters of traditional BaNtu people and sunbathe in African ascendancy…" (2023, p. back cover). I feel the flow of ancient wisdom seeping through these pages as a contribution to the accumulation of knowledge to extend growth and human development.

Nobles has studied BaNtu cosmology and a summary related to KABAFA is fascinating. KABAFA is basically a reformulation of this ancient wisdom, so there is no claim that the conceptualization is entirely original because our ancestors have already thought these thoughts before.

I have discussed the science of melanin for decades, and I have claimed melanin to be the conduit between the material and spiritual realms of existence. A more sophisticated discussion is required to fully place this complex biopolymer into proper context. For a Grand African Narrative on the science of spirit, the term melanin is not just classified as carbon element or a major element required for life, it is a critical molecule of life that serves as the energy source for our KABAFA.

By studying under the Congolese ancestor, Baba Fu Kiau Bunseki, Nobles interprets from this master BaNtuologist. In a summary of Bantu-Kongo cosmology, Fu Kiau (2001) explains that a heated force of kalunga blew up and down as a huge storm of projectiles that fuse together and in the process of cooling, solidification of the fused mass occurred, giving birth to the earth. It is believed that all of reality (kalunga) is a process of perpetual and mutual sending and receiving of spirit (energy) in the form of waves and radiations.

According to the Bantu-Kongo cosmology, the human being is a living sun (i.e., energy). As we ponder the universe, we possess the capacity to know and seek knowledge, and the spirit of a man/woman is the energy to keep us in a perpetual relationship to perceive and ponder the universe. In contemporary times, we have oversimplified this ancient concept by relating it to "melanin."

As stated by Nobles, the Bantu-Kongo believe that diverse forces and waves of energy govern life and surround humans.

This fire force called kalunga is complete in and of itself and emerges within the emptiness or nothingness and becomes the source of life on earth (Nobles, 2023, p. 58). From kalunga to fire atoms to this fire force, melanin is what we have called this science in modern times. If we follow the transgression of information to create an African Grand Narrative as suggested by Nobles, this "would assert that human beings as Spirit Beings are those who live and move within and are inseparable from the ocean of waves or radiations of spirit (energy)" (Nobles, 2023, p. 60). To begin with a redefinition of melanin, the following chapters will break down the use of KABAFA.

CHAPTER 2

The KA

Fig. 2 For KA, the arms raised symbolize the surrender to the almighty creator.

The Metu Neter symbol for KA is represented as two raised arms. KA is the first the division of the soul as described by ancient African scientists. The two arms raised are a surrender to the almighty creator, but the representation of the arms makes reference to the body. The body becomes the vehicle that houses the soul and with no physical body there is no entity to capture the essence of creation. The body must surrender to the cosmic elements to manifest the awesome power of the universe. Without the body, there is no soul to discuss. The KA, therefore, surrenders to the universe to capture the spiritual material connection.

The darkness of the body is important to capture the spiritual elements that exist in the universe. In other words, the pigmented body is constructed to absorb all the electromagnetic sources of energy that exist. As described in the book, Dark Matters Dark Secrets, the melanocyte in the skin of the body is like the cellular KA. The arms of the melanocyte are raised like the arms in the Metu Neter representation of two arms raised. The sun bathes the skin and the melanocyte cells are extended to receive this light and cause the melanin to rise and increase the black-light experience.

Fig. 3 The cellular KA is the melanocyte in the skin strategically positioned to receive light.

As we describe KA as the First Division in our word KABAFA, the KA becomes the basic unit in which we find the foundation for seeking the spiritual connection to the unseen world. The darker the cellular KA, the more connected the individual will be to spiritual and unseen elements. We need to have KA to be able to fully manifest the power of pigmentation.

According to the interpretation of Akbar (1994), the whole notion of KA is viewed as being the body of the person - one's physicality. Without the construction of the pigmented body, there would be nothing to synthesize the spiritual energy. When we say ashes to ashes and dust to dust, we are referring to the human body and to its connection to earth and the cosmos. Since we are a microcosm of the macrocosm, we can easily visualize the importance of the KA as the basis of the soul. Having no body equates to having no soul.

A person cannot transcend from a lower level of consciousness to a higher level without the body being in perfect alignment. Therefore, pigmentation in the skin becomes an essential element to absorb the electromagnetic energy that is in the wavelength containing spiritual energy.

From cosmic energy down to radio waves, the physics of melanin plays a key role in absorbing energy. The darkness in the KA is connected to the energy emitted from the electromagnetic spectrum. It becomes difficult to feel the vibe of any spirit without having a body with hue that is ready to receive the ecstatic energy. Albinos, however, who are genetically, consciously, and culturally connected to an African heritage, are an exception.

What is meant when James Brown screamed and Eric B. and Rakim spewed, "You got soul; If not you would not be in here." Some people feel it and some people do not and this is directly correlated with a person's KA. The darker the KA, the greater the connection to the soul. From music to any physical element harnessing energy, dark bodies feel the intensity immensely. The KA is the foundation for our redefinition of melanin as the critical element in defining the soul. It is the body - next is the breath.

CHAPTER 3

The BA

Fig. 4 For BA, the human-headed bird symbolizes the breath of the soul.

When thinking of BA, it is represented as a human-headed bird. As the second division of the soul, the BA combines with the KA to generate life, and without breath, there is no consciousness, and the soul has then transmigrated into another realm.

Breath is dependent upon oxygen as an essential element. The darkness of melanin in and outside of the body absorbs oxygen. Too much oxygen can be toxic, so a pigmented body must be adjusted to prevent oxygen toxicity. Oxygen is the gas for the engine that propels breath. Breath is the only power controlling your existence, so the BA becomes critical for life. When a person's breath ceases, the soul has moved to another realm of existence.

As described by Akbar (1994), breath is the invisible energy source and vital principle which represents the essence of all things. When a person physically expires and dies, the last breath is when the soul enters another realm of consciousness.

It is interesting that melanin serves as an effective natural antioxidant. Therefore, it becomes important in the manifestation of breath and the control of oxygenated states.

The BA of any person is the life source of the body, and the presence of melanin fully equips the body with the proper channeling of energy. People who meditate and control their breathing are channeling oxygen through pigmented areas in the nasal passages and internal organs. Melanin is a neutralizer of toxic substances, and it prevents oxygen from damaging cells. In a person's breath is the inhale and exhale of a cosmic consciousness. There is no state of consciousness that exists without breath, so BA is critical to our understanding of soul studies.

The biopolymer properties of melanin allow pigmented cells to stay healthy. The better the quality of air that a person can maintain increases the likelihood of a pleasant outlook in life.

Most people are stressed because they do not pay attention to their breathing. Slowing down and preventing hyperventilation can increase consciousness. In effect, a person's soul is prepared for growth and development under conditions in which breathing is properly managed.

In the natural process of life, people normally die when they are older and aging. Melanin in a pigmented body reduces aging to keep the KA alive. However, there are consequences from oxygen toxicity or free radical exposure (Papas, 1999).
For example, there can be:

a. Somatic mutations
b. Loss of enzyme expression
c. Cellular damage to microtubules
d. Decreased integrity of subcellular organelles
e. Impaired nutrient diffusion
f. Impaired tissue viability

For a scientific understanding, free radicals formed from oxygenation could be superoxide, peroxyl and hydroxyl. The enzymes that can work along with melanin to neutralize those free radicals are superoxide dismutase, glutathione, peroxidase and catalase.

In sum, BA is critical to life and pigmented cells help regulate the use of oxygen for the KA to be alive and healthy. As the second division of the soul, BA is the driving force for your existence on planet earth. You leave earth and enter the spiritual world permanently when your breath ceases to manifest. However, your soul is in existence for eternity. Calling on your ancestors evokes the souls of those who came before you and their souls continually impact you in your remembrance.

CHAPTER 4

The FA

Fig. 5 For FA, the carbon-based molecule is the essential element creating all life on the planet and it exists everywhere.

This chapter validates the essence of why this new formulation of melanin as a term is being reconsidered and renamed. There is no way that we have a monopoly on knowledge today with the enormous amount of ancient and historical information that is still being uncovered. Without the body (KA) and the breath (BA), there are no cells to function. The FA are the fire atoms (i.e., FA) where the cells function.

Collectively, KA, BA and FA give us our soul experience. The FA represents the words fire atoms. Fire atoms are explained in the book called Stolen Legacy by George G.M. James. James first published his book in 1954. He was writing during a time when it was difficult to express anything positive about African/Black people in North America. He took a deep dive into Greek philosophy and uncovered how it is a stolen Egyptian philosophy. James provides detailed explanation with a wealth of resources to prove his point about the origins of these philosophical terms. In other words, the Egyptian mystery system is the source of where the Greek authors gained their knowledge.

Due to the written record, there are names that have been associated with this philosophical discussion on fire atoms, and Democritus is one such name. According to James, the dates of Democritus's life are from 420 to 316 BC. Checking other sources obtained on Wikipedia, the life Democritus was from 460-370 BC. Interestingly, Leucippus is traditionally credited as the founder of atomism and it is thought that he developed it with his student, Democritus.

What makes the story more confusing is that Diogenes Laertius claims that Plato disliked Democritus so much that he wished to have all his books burned. However, Aristotle was not as harsh towards this magician. As we say confusing, we can also say fascinating because of the time period of Aristotle (356- 323 BC) and Alexander the Great (356- 323 BC) destroying the libraries in Kemet around 332 BC. With this invasion, the royal temples and libraries were plundered and pillaged, and Aristotle school converted the library at Alexandria into a Research Center.

Aristotle taught Alexander and Democritus was an influential Greek person that was respected by Aristotle. Some reports indicate that Aristotle wasn't involved in the death of Alexander. Listen, James is right on target when he suggests that there is no way possible that either Democritus or Aristotle could have produced this large number of books by either of these men in a single lifetime. For Democritus, more than 60 different subjects are treated, and the subjects include Ethics, Physics, Astronomy, Botany, Zoology, Poetry, Medicine, Dialectics, Military Science, Law and Magic.

It is thought that magic is what Democritus was first known to be associated with. What James did was ask the poignant questions, like "How did Democritus accumulate those books if he did not write them?" see James pp. 78-79.

According to James, Democrats taught nothing new, but he simply expounded or copied what he learned from the African/ Black creators of this knowledge. As we link his connection to magic, his doctrine on the universal distribution of fire atoms is based upon a magical principle: if the atom is an ingredient of the world, then it would be universally distributed. With the historical evidence that Democritus enters history as a magician who visited Kemet, he studied as a neophyte and was trained in concepts he probably could not personally comprehend or feel the true magic. In fact, building pyramids and colossal structures might have required magic that we do not understand today.

Lastly, observe the depths of the disconnect from the African/Black philosophy because of the intellectual turmoil created in Greek society. The age of Greek philosophy was 640 to 322 BC. Democritus was a pre - Socratic Ionian philosopher and the life and times of he and Leucippus are sketchy. However, later in Greek history, Socrates was persecuted by his own people for teaching a strange doctrine, and Aristotle in his waning years of existence was fearful of having the same fate. This Greek philosophy they were spewing was not authentic and even the collective Greek society during their time period in history had concerns. Concerns enough to kill these "great" men.

I will first repeat how James records the history on these concepts and, secondly, I will give the modern day parallel to the metaphysics of melanin. The goal is to link these ancient conceptualizations to a present-day view of the physics of spirituality and the soul with pigmentation as the central theme. According to ancient texts, the mind (nous), no matter what other name or function we give it, is fire, since it is composed of fire atoms, which underlie creation. Now we know all creation comes from the mind and a spark that lights the fire for creation.

The mind or soul is composed of fire atoms, which are the finest, the smoothest, and the most mobile.

These fire atoms are distributed throughout the whole universe; and in all animate things, and especially in the human body, where they are found in the largest numbers.

These fire atoms are the igniting elements that serve as the conduit between the material and spiritual realms of existence. Material can be touched whereas spirit is formless. Even though the physical elements are different, they both contain atoms of fire. To make a distinction between material and spiritual or life and death, the experience is due to an arrangement in the atoms. In the atomist view as explained by James, when the atoms are arranged in a certain way, life emerges; but when that arrangement is changed to another way, then death is the result.

In death, the personality disappears, the senses also disappear; but the atom lives on forever. The heavier atoms descend to the earth; but the soul atoms, which are composed of fire, ascend to the celestial regions, whence they came.

So when you die, your physical body is here on earth as a buried corpse or burnt into ashes that are still on earth as used fire atoms. The spirit, however, rises and continues to manifest with its lighter fire atoms in perpetuity. As we deal with emotions and feelings, it is interesting how the fire atoms emit minute images on physical objects that impress upon our senses (like auras) which set in motion our soul atoms (spirit or vibes) to create the sensation and knowledge to connect with nature and experience life to the fullest. On a cultural level, this explanation is why many heavily pigmented people are more in harmony with nature verses in conflict and seeking control of nature in many less pigmented people.

As the last component in KABAFA, the fire atoms (FA) are directly related to the metaphysics of pigmentation in the human experience. For this author, the ancient discourse on fire atoms reveals to me that the Greek derivation known as melanin is symbolically fire atoms. The Greeks only defined melas as black, and we Eurocentrically use melanin to define "Black" people. So of course, the question becomes,

"What is Black people in the hueman experience?" With all the different shades, colors and pigments in people who are melanin dominant, how can a people really be classified as Black people? What is really described is a people who are culturally, consciously, and spiritually connected through a soul experience we call KABAFA.

CHAPTER 5

The Kemetic Mystery Revealed

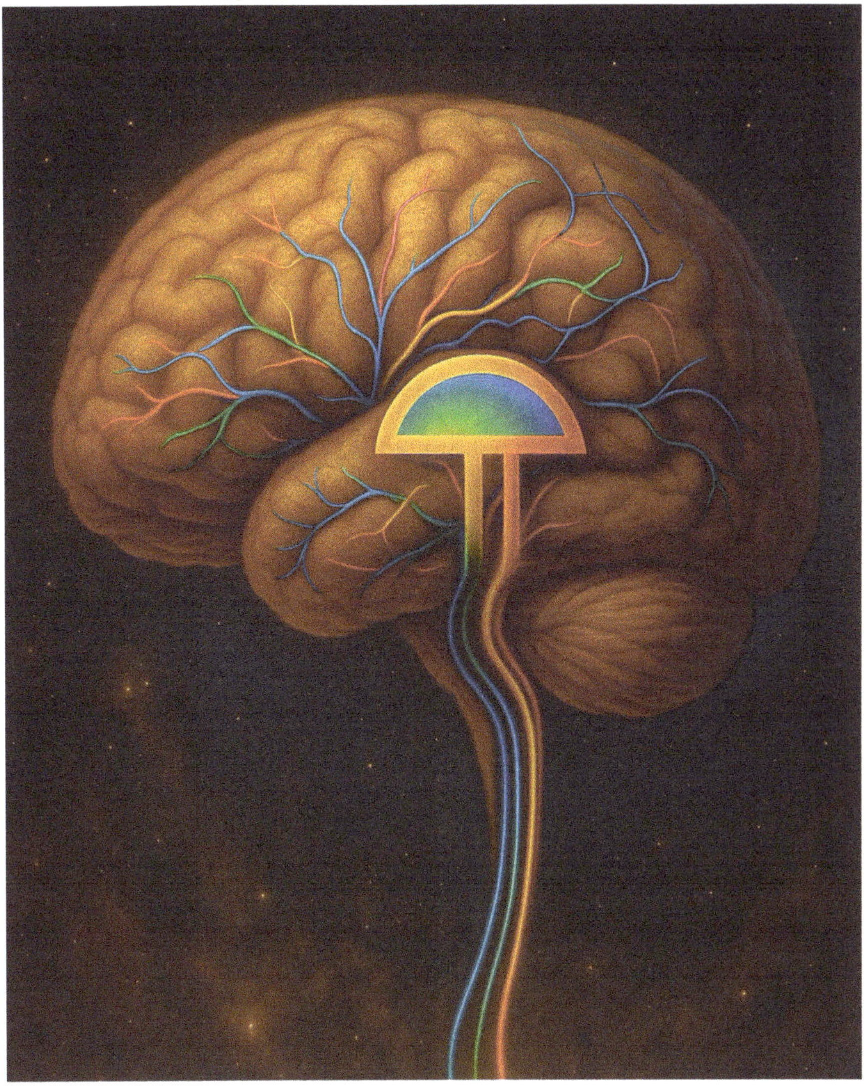

Fig. 6 The brain-stem is at the core of our consciousness, and this is where the Amenta nerve tract resides.

Fig. 7 Ancient symbol of Amenta, representing the subconscious realm and gateway between life, death, and rebirth-anchored at the brain's base and extending into the spine.

The Kemetic Mystery System was emphasized in the work of George G.M. James. James spoke of a system where the written word was limited to halt the theft of sensitive knowledge. Without writing, there is no theft from a foreign intruder, and secrets are maintained. Unfortunately, those with ulterior motives who are learning the new material can write it down and claim it as their own.

The current day Masons and Secret Societies have knowledge and information that is not revealed to the public. In modern history, the Vatican is a location in Italy in which the Roman Empire hoarded ancient knowledge in a five-mile underground location. The Romans came after the Greeks and the Greeks were taught by the people of Kemet, specifically the priest in Kemet. As revealed by James in the Stolen Legacy, according to Herodotus, the Egyptian priests possessed supernatural powers, for they had been trained in the esoteric philosophy of the greater mysteries and were experts in magic.

Fig. 8 TiMo in front of the Temple of Karnak, one of many places in Kemet in which foreigners came to Africa to learn.

Fig. 9 Black statue found in the tomb of King Tutankhamun.

Fig. 10 Image of people in ancient Kemet using the ankh as a symbol for energy.

If you read The Egyptian Book of the Dead, it is a book of magical formula and instructions, intended to direct the fate of the departed soul.

It is the spirit we are and the soul we have that allows us to reach both the divine universal consciousness and the place where the departed souls go. While alive with Ba (the breath), the mystery of life is to move into this universal space to become God-like via the summum bonum, which is the supreme good in man to become God-like. The soul is the result of the spirit-body union and our new term KABAFA describes the connection between the body, the breath, and the fire atoms that provides the spark for this divine universal consciousness. Man's ultimate goal consists of achieving the marriage between the soul and spirit (Baines, 2008), and this merger is not mastered by many.

The revelation in the mystery being revealed is that we all have the capability of being God-like. It requires dedication and study to be the best that you can be. Nobody has a monopoly on knowledge, but there are people and organizations that do try to control and hoard information to their benefit. This has happened historically, but we must be sophisticated to understand our role in this experience called being a human. KABAFA is a start as we re-conceptualize ancient knowledge for current advancements in the human experience.

Most of the time the reason why knowledge is held in secrecy is when someone is trying to control the information to their benefit. Seldom do we think of secrets being held in order to diminish access to get into the hands of the wrong people. Think about your world when you have invaders who come into your community, and you feel a loss of your culture. You may not be xenophobic and want to kill someone, but you want to preserve your culture. The modern impact of gentrification is a good example of an environment where a mystery system could be developed. Your community is being gentrified by a foreign element, but you can save your sanity and preserve your culture by developing a system to maintain your historical records. In reality, gentrification is economically suppressing, so a mystery system might not halt the onslaught. As an example, we can historically say that a Kemetic Mystery System did not prevent the onslaught from invaders into the established culture of Kemet. The intellectual community was destroyed, and then the esoteric knowledge could only be preserved and passed on to those who are privileged to know.

By watching and listening to the Daily News, you have seen the intensity of race relations today boiling over into intense verbal and intellectual arguments. It is silly to entertain the thought that the paternalistic mindset of Eurocentric thinkers can come to an agreement with formally oppressed people when thinking about true freedom. The integration versus segregation debate is a paternalistic Eurocentric approach but the revolutionary non-Eurocentric approach is liberation versus control. Freedom cannot become obtained with either integration or segregation, but freedom can be obtained with liberation.

Liberating our souls from the fetters of this human body is the crux of the Kemetic Mystery System. There is a curriculum and social order needed for liberation, and magic was a specialization that was mastered.

As revealed by James in his review of Herodotus who lived nearer to the time of the Egyptian Mystery teaching, "The Egyptian priests possessed supernatural powers for they had been trained in the esoteric philosophy of the greater mysteries and were experts in magic. They had the power of controlling the minds of men (hypnosis), the power of predicting the future (prophecy) and the power over nature (i.e., the power of Gods) by giving commands in the name of the divinity and accomplishing great deeds" p. 134.

The Dark Universe and Amenta

A current mystery in science is dark matter and dark energy as scientific concepts. Scientists know dark matter and dark energy are out there in the dark universe, but modern scientists are having difficulty explaining it. Let us explore and relate this modern mystery to the Kemetic Mystery System.

From a gentle read of two profound mental health scientists (i.e., Richard D. King and Edward B. Bynum), we can see the Kemetic Mystery System revealed in their breakdown of the Amenta world. Dr. King first explored the concept of the Black Dot as the doorway to the unconscious (King, 1990), and he expanded our understanding of the ancient term called Amenta by linking it to the neuromelanin in our brain stem (King, 1995). Dr. Bynum (2012) goes even further in his book Dark Light Consciousness to explain how the spine, brain stem and deep brain core structures containing the bioluminescent core for our link to the unconscious world. Although Amenta is associated with death, it is an ancient term that connects to this unconscious reality described by King and Bynum and in the dark universe. The Amenta nerve tract is the physiological link to both the unconscious realms of existence and the gateway to the afterlife.

The dark universe is really the cosmic Amenta. The darkness of the universe captures light, and it does not let it go. Like that black shirt on a hot day, light isabsorbed, and you feel the radiation. The dark universe is made up of equal parts of dark matter and dark energy. Current science has demonstrated that darkness pervades the majority of the universe so we must give credence to the interplay between dark matter and dark energy.

Dark matter pulls things together - it attracts. Dark energy pulls things apart - it repulses. From a Eurocentric perspective you hear astrophysicists explain in warlike term the opposition of these forces and how this opposition dictates the discussion. Rather than having dichotomous logicand splitting it up, an African-centered way of viewing this is to describe this so- called opposition as a harmonious brew of cosmic juice that creates life in the seen and unseen world. Therefore, if we think from an African-centered perspective, you see a harmonious relationship rather than thinking of opposition.

This harmonious relationship began 13.8 billion years ago when the universe was first created. Scientists have since thought and confirmed that matter cannot be created or destroyed, that is a general scientific term in Newtonian physics. On the other hand, dark matter can be some kind of matter (e.g., fire atoms), but it does not interact with light, so we do not see it. Protons, electrons, neutrons and quarks may be observed to make atoms work to subsequently make molecules exist, but with dark matter, the elements have not been detected by modern science yet.

Similarly dark energy is found in empty space so you might expect it to be nothing, but it is an energy that exists in the openness of space. It is not too far-fetched to think this openness of space is the Amenta or Netherworld where the ancestors dwell.

Amenta is an ancient African term that represents the land of the dead, the underworld. It became a symbol of Duat, the realm where the dead dwelt. It is a significant part of the cosmology in Kemet and Amenta represents the river Nile, the dead, the afterlife, and the sun. The western region of the Nile became associated with the dead, where the sun set every day. There are depictions for the Land of Amenta or the Desert of Amenta, and we know there are important burial sites in Kemet on the west bank of the Nile River.

Our ancestors lived for the afterlife and death was just an interruption in the passage of time. The ideas of immortality and a world after death demonstrates their reasoning that existence would continue after the KA ceased and the BA no longer had breath, but the FA was able to continue to manifest for eternity.

As deep as many astrophysicists contemplate the existence of dark matter and dark energy today, ancient African soul scientists contemplated divisions of the soul and ordered them into distinct divisions. An example can be observed in Table 5A below.

Table 5A – The Seven Divisions of the Soul

Ka	The physical body of the person
Ba	The breath of the soul
Khaba	The shade or covering of the soul
Akhu	The psychic nature of the soul
Seb	The self-creative power of being
Putah	The ability to reproduce intellectually
Atmu	The divine soul of perpetual continuation

Think about how your energy can get zapped by a comment, phrase or external cue that alters your reality at the instantaneous moment. Your soul has shifted, and you no longer feel like you did a moment ago. By breaking down divisions of the soul, it is not like your personality has changed, it is your soul that has shifted. We need to give greater exploration to the ways in which we can explore the soul.

CHAPTER 6

Soul Studies

Fig. 11 The human body is the vehicle to transport us to higher realms of existence.

The contemporary voices that are delving into soul studies are numerous. To stay true to the field of studying spiritual technology and the manifestation of the soul from a pigment-powered orientation, further readings by Tolokun (2020), Nelson (2003, 2014), Carson (2019) and Menzise (2015) can be explored by the reader. It is important to expose you to the thinkers you have not heard of in the "New Age" exploration.

In many Eurocentric bookstores, there is a category called New Age. New Age is a category of material discussing ancient principles and concepts revised for the Eurocentric frequency. The deciphering of the ancient knowledge is far from being new but is a way to monopolize knowledge and extract any Black/African connection to the base of this knowledge. Along with this current text, it is important to cite literature by African-centered scholars.

Menzise has a clinical background, and he provides a thorough analysis of initiation rights and explorations into spiritual systems. He emphasizes that the spiritual and initiation legacy presented by James in his book *Stolen*

Legacy is critical as a foundation for the cultural, political, intellectual, and all other accomplishments to make civilization possible. Therefore, study Menzise since he is an initiate into African spiritual systems.

Tolokun is trained in African spiritual systems, and her counseling background has allowed her to focus on psychospiritual development from a traditional African perspective. Study Tolokun as she demonstrates the connection between the ancient Kemetic discipline of Sakhusheti (soul science), the contemporary field of Afrikana psychology, and her approach to Afrikana spiritual coaching.

Nelson has an overlapping approach to what is expressed as KABAFA in this text. In fact, there is slight variation of the ancient terms for she uses Ka Ab Ba for her exploration into the Tree of Life. As she is helping others to self-actualize their potential, Nelson delves deep into the MerKaBa which is similar to what is meant by KABAFA. MerKaBa is a luminous body of light. It is your energy body, and this is where KABAFA is ignited. For Nelson, MerKaBa is the physical form of the human body. As a vehicle MerKaBa will help the individual be aware of and travel through the journey of consciousness, rebirth, transformation, resurrection, ascension, transport, empowerment, enlightenment, initiation and service to humanity.

Carson also mentions MerKaBah as the vehicle for ascension for the enlightened person. According to Carson, the fire in the middle of MerKaBah is the naval chakra or sacral plexus near the naval area.

What is even more valuable to this discussion on KABAFA is the saying of THOTH from the *Emerald Tablets* that are interpreted by Carson.

From the Emerald Tablets 3 THOTH said,

"Fire, the inner fire
is the most potent of all force,
for it overcometh all things and
penetrates to all things of the Earth.
Man supports himself only on that which
resists.
So Earth must resist man else he existeth not.

All eyes do not see with the same vision,
For to one an object appears of
one form and color
and to a different eye of another.
So also infinite fire,
Changing from color to color,
Is never the same from day to day.

Fig. 12 Thoth is represented as the God of Wisdom.

Essentially, the pigmentation of melanin makes people see and feel things differently. As expressed, thousands of years ago, these fire atoms are everywhere and with fire being a potent force, people have different experiences in Soul Studies. As we look at the impact of the medical-pharmaceutical-industrial complex, we see how natural plant medicines as well as synthetic medicines can have ethnospecific variations in their effects. The capitalistic corporate system must keep you hooked on synthetic medications in order to keep you in the "system." You become a threat to the system when you devise ways to stay healthy without needing the medical-pharmaceutical-industrial complex. These synthetic medications can be damaging, and human beings have been controlled and manipulated away from seeking answers in nature.

The evidence to reveal the full plan for destroying your health is when you visit the doctor's office for the first time. The basic intake question is, "What pharmacy do you use?" It is as if the plan is to get you hooked and latched on to medications. The irony is the Food and Drug Administration (FDA) is non supportive of natural remedies, and this is a joke because any organization or administration with food and drugs cannot be for your safety. Both food and drugs are controlled by major corporations, and corporations and the Chief Executive Officers are only interested in stock increases and money. It is quite interesting or revealing how doctors never ask, "How is your spirit?" or "How is your soul?"

Therefore, we need to engage our thought towards studying the soul. To be truthful, this is not simple to investigate. It is difficult to study something you cannot touch. Unless you have the correct instruments, the reality of manipulating unseen phenomena can be impossible. The novice would think this is difficult to grapple with, but for those trained in clairvoyant and psychic abilities, they hold the evidence for delving into the spiritual state of the soul.

If spirit is energy, you can actually see a person and their energy change when they are manipulated by external factors. These external factors, for example, could be from the television, the radio, the voice from an acquaintance or memory of a past event. By watching a person's energy change from these external factors, we are observing spirit or soul of a person being changed. Beyond the external factors, there are obviously internal dynamics that can also have an impact the person's energy which is displayed as their spirit or soul. What you put into your body by ingesting, hearing, smelling or consciously perceiving an event is what can impact your spirit. With these external and internal factors, there are a host of situations that can be presented to study the soul. We are saying that external and internal factors that can impact the body, mind, and consciousness are the evidence that the soul can be studied.

Fig. 13 TiMo in Cali, Colombia and the Andes Mountains searching for medicinal plants.

Plant Medicines

The opportunity to explore the inner consciousness of reality was sought in the seclusion of the Andes mountains in

South America. Far away from the influence of technology and urban distractions, this was an exploration into the realms in which multi-universes are connected. In the age of 5G technology pollution, it was interesting to tap into an energy that pervades the universe. The naked eye can overlook these subtle levels of existence. Therefore, people will never be able to tap into the universal consciousness, even with their eyes wide open. The key is to tune into the "third eye" that was born and developed for us to space travel.

Dreams during sleep are the closest phenomenon to explaining the non-locality of perception. In other words, when you dream, you can have events that are totally disconnected with time. A childhood event could be merged with a current event in your life as well as people, places and times that strangely overlap. Our existence as human beings can be recorded in ancient records called the Akashic records. These events are thought to connect all of humanity. The inner workings of the brain are the foundation to tap into those ancient records, cosmic events and answers to where the soul and spirit dwell.

I am trained as a scientist, so it is difficult for me to exclude my science background from an analysis of the spirit and soul investigation. Personally, I feel balanced in my life, and I am not disrupted by any blocks or past traumatic experiences that I can consciously recollect. I feel mentally, physically and spiritually fulfilled. In the Sixth decade of life, I am only managing the prostate challenges experienced by many African American men. I do have benign prostatic hyperplagia and elevated prostate specific antigen levels. I took the internal changes seriously, and I proceeded with the established route and had a biopsy and MRI and there is no cancer detected. If I am seeking healing or answers from pursuing plant medicines, then healing my prostate functioning or finding clarity on my next journey in life will be exciting. If there was any trepidation, it was the offsetting of the balance I feel in life. To be positive, I sought plant medicines to seek the medicinal powers, to fully enhance my life and to study the soul or spirit.

Many people call this the Ayahuasca journey and you can read and view the various experiences people have had from all over the world. Rather than me seeking something, many people say, "the vine calls you." I believe this because while cutting grass, the kudzu growing in my backyard intimately revealed to me how this vine "thinks" to intertwine and latch on to the fence, other trees and bushes.

Ayahuasca is a psychoactive drink containing stem bark of the Banisteriopsis caap vine in combination with leaves of the Psychotria virdish bush. N,N Dimethyltryptamine (DMT) is the actual stimulant to raise the inner awareness, and it has been called the spirit molecule.

Fig. 14 The dimethyl tryptamine molecule (DMT) is one of many spirit molecules.

There are many natural psychoactive compounds (Cannabis, Peyote, Mescaline, etc.), but I experienced Ayahuasca which some call the “vine of the soul.” Ayahuasca is a traditional Amazonian decoction that actually has many names. As previously mentioned, the psychoactive drink is made from the stem bark of the banisteriopsis caap vine, rich in beta carboline harmala alkaloids, usually in combination with DMT-containing leaves of the Psychotria virdish bush. DMT can affect pineal gland functioning and stimulate certain receptors.

Interestingly, DMT can be found in human urine, blood, and cerebral spinal fluid. DMT is likely to serve as an endogenous sigma-1 receptor ligand. Although not well understood, this ligand can be found in the prostate, lung, ovaries, breast, liver, and brain. The suggestion is this receptor may play a role in depression, anxiety, and cancer. I do not have any of those conditions, so I did not know what the connection to healing there may have been. Scientifically, there are a multitude of other physiological effects, so there is potentially no limit to how DMT is raising consciousness to elevate us for the connection to the soul of the individual. As the pun is intended, the acronym DMT for me means Do Moore Tim.

With a variety of possible medical uses that require more research, the use of Ayahuasca dates back to the earliest aboriginal inhabitants of the Amazonian basin. It was used by indigenous shamans for communication with spirits, magical experiences, rites of initiation, and healing purposes. In Table 6-A, there is a synopsis of the plant medicines that were involved in the process. In addition, Table 6-B is a brief list of the endogenous neurotransmitters that are linked to how the plant medicines affect consciousness.

Table 6-A
Sample Plant Medicines and Usage

Medicine	Usage / Notes
Rappe or Rape (pronounced hapeh)	Finely ground tobacco and other herbs blown up the nose. Nicotiana tabacum activates acetylcholine, dopamine, and epinephrine to boost the immune system and cleanse chakras.
Sanaga	Eye drops to sharpen vision. Made from the roots and bark of a shrub called Tabernaemontana undulata.
San Pedro	Also known as Quechua; a cactus used for psychedelic natural medicine and healing.
Ambil	A paste, jelly, or syrup made from tobacco and other organic plants and herbs placed in teeth or gums to cleanse, clear thoughts, and open heart and throat chakras. Connects to spirit with a very masculine energy.
Mambe (Ypadu)	A fine powder made from toasted coca leaves mixed with an alkaline reagent, often yarumo (Cecropia peltata) ash, quinoa ash, or burnt crushed seashell. Mambe means "tongue of God" or "word of life," and has a feminine energy.
Ayahuasca	A psychoactive drink containing stem bark of the Banisteriopsis caapi vine in combination with leaves of the Psychotria viridis bush.

Table 6-B
Endogenous Spiritual Enhancers

Compound	Notes / Role
Tryptophan	Amino acid, precursor to serotonin and melatonin
5-hydroxytryptophan	Intermediate metabolite in serotonin synthesis
5-hydroxytryptamine (serotonin)	Neurotransmitter regulating mood, sleep, and cognition
N-acetylserotonin	Intermediate in melatonin biosynthesis; antioxidant role
5-methoxytryptamine (melatonin)	Hormone regulating circadian rhythm and sleep

For my experience the shaman who prepared the drink started preparation in July 2023 and three months later it was ready for access into the deep realms of consciousness. Since I was born in July, I feel a connection to the preparation. To be honest, I went into the experience freely without any expectations. It was a journey to explore the full experience of what has been prepared for this book on KABAFA.

Although I personally experienced no psycho genic stimulation that was overwhelming to my consciousness, I observed evidence of the spirit being manifested on the physical plane of existence. Over a two-day period, a trained shaman (named Suecha) guided all participants on the journey. Shaman Suecha prepared all medicines to ingest, delivered all incantations, and served as the conduit for the spirits that were evoked. Each person had a very different journey into the underworld, and this is evidence to me that the person's established consciousness has everything to do with their spiritual elevation.

Rather than say elevation, the correct term should be manifestation. In this experience, Suecha spoke Spanish, so it was not coherent to feel the full language for English speaking participants. I had a separate cleanse during the first session because Suecha felt as though there was an energy block for me. The only psychogenic experience I felt with my eyes closed is seeing the waters of NUN flowing first and then the celestial stars of the goddess NUT. Since I am African centered in my thinking, that is the experience I describe. My stomach and bowels were stimulated, and a bowel release happened several hours later after session one. Certainly, this was not a great feeling.

We know the brain and gut are connected, so the plant medicine works on both the nervous system and the digestive system. After the first session, Suecha did a separate cleanse consisting of applying herbs to my upper torso and chanting around my body while seated. Through an interpreter, it was revealed through Suecha to me that a deceased relative was happy I attended. In addition, I was given blessings on my "project." Suecha never met me before this spiritual event, and he did not speak English. Therefore, it was confirmation to me that the spirit was flowing for me. My stepbrother passed away two months earlier and my intentions for seeking Ayahuasca were for good health and guidance on my research projects and for direction in the next decade of life. It was revealed to me by Suecha that my energy was balanced, and I was protected.

Now, although I did not have any severe reactions or deep experiences, my fellow participants felt deep expressions of healing that were manifested through intense crying for some and joyful laughter for others. I saw it expressed in those around me. A part of the Ayahuasca experience is the uncontrollable urge to regurgitate. This purging is to release blockages. It does make you feel nauseated, but the outcome is a spiritual awakening. I saw it.

I need not go into detail about the personal experiences of my fellow participants, but I did observe the spirit jump from person to person while around a campfire during those hours after midnight when the spirit feels freer to lurk and appear. After many people regurgitated, that is when the intense expressions were displayed and each time, the Shaman absorbed the experience of the person going through the experience. It was quite evident this was physically taxing on the Shaman.

I did not feel any entity speak directly to me or evoke a feeling in me, so that is my personal journey. However, during session two, when I was prepared to depart and go to my room and rest, I touched my colleague on the shoulder to ask if she was OK. I was prepared to leave and seconds after touching her, I had the uncontrollable urge to violently regurgitate for the first time. Meanwhile, others had this reaction several times over the two-day session. In sum, I experienced the spirit of another person's experience jump to me as if it was not finished expressing itself on the physical plane. Many felt an awakening from this journey, and I can testify that I saw the spirit and soul come out and play.

In this South American experience in the rain forest of the Andes Mountains, the participants taking the plant medicine were both melanin-dominant and melanin-recessive. Since everyone was a human, the spirit worked through everyone's KABAFA, but the experience was different. With KABAFA serving as the conduit there are going to be different experiences. Concisely, spirit can manifest through any biological entity, but the expression of the soul has a more complex display. If a nation can have a soul and nations can be good or bad in their intentions, it is the same with people and the expression of the soul.

The history and past experience of a nation defines its soul. Similarly, people have past experiences of being righteous or evil, and this molds the soul. As described in the chapter on the Kemetic Mystery System, there are different aspects of the soul, and some divisions go into detail about moral and intellectual development. The indigenous people in the Amazonian basin who have traditionally used plant medicines are balanced with nature, so there is a clear soulful expression. In other regions of the world, different plant medicines are used for a similar journey (e.g., Tabernantha Iboga). In central Africa, this plant medicine is known as Ibogaine to treat addictions.

Near Death Experiences

Near death experiences are another realm of investigation that can shed information on the presence of the spirit. The spirit could leave the body and reside in a location to observe the physical body going through a temporary maladjustment. It could be a car accident, cardiac arrest, drowning or any situation where breath has ceased, and the heart is not pumping blood. Many people have died suddenly, but there are occasions where others are present to revive the body.

It is interesting that during cardiac arrest, electrical boosts must be administered along with CPR. The body is electric, so KABAFA needs extra stimulation to keep the person physically alive. In soul studies, a component of this discipline would be what is happening when people have a near death experience (NDE).

The mind or consciousness elevates outside the body and the person who is revived often speaks about a hovering experience over the body. The hovering is the spirit waiting for instructions on the next journey - to stay or go? We know which brain areas are linked to spirituality, so it is possible to associate the brain as the command center for the spirit to be expressed through KABAFA.

Cosmology and Culture

In soul studies, it is critical to take into account the culture and cosmology of the person making assessments as well as the recipient of the information. Beyond differences in the expression of KABAFA, the mental state and what a person believes is the crux of the matter. There are cultures that desire to control nature and there are other cultures that live in peace and harmony with nature. Therefore, the understanding of spirit and soul depends on your worldview.

A critical component of studying these metaphysical concepts is to be grounded in nature and understanding the connection to nature. Only when people are taught in this manner will we ever be accepting of things we cannot physically see. Our eyeballs actually blind us to the reality of our spiritual nature. We must develop the skills to open our third eyes to be fully engaged in soul studies. KABAFA is a term that can be used to set the tone for investigations that are culturally competent and cosmologically relevant. Our spiritual connection is dependent on holistic thinking and a balanced outlook on life.

Soul Therapy

The first prayer - *Now I lay me down to sleep I pray the lord my soul to keep, if I should die before I wake, I pray to the lord my soul to take.*

Soul Therapy or Spiritherapy are defined as the treatments to help nurture a person's spirit and soul in the proper direction. The proper direction is in the positive direction, and this requires positive thinking. This is not to describe this as positive psychology, which is a course in the discipline of psychology. The difference between Positive Psychology and Soul Therapy is the emphasis on a person's inner space.

For example, too much is emphasized in this capitalistic society on material gains. The gains are to acquire things from outer space; that is items outside of the self. To focus on soul therapy, a therapist must guide the person internally to the place in their consciousness where smiles are generated. At the core of people's personalities, there are simple things that make people happy. Here on the internal landscape of the mind are the answers to spiritual elevation, heightened emotional responses, and higher states of consciousness.

For meditation, sometimes the hardest aspect of meditating is leaving the outside - outside. People need to be trained to master the semi conductive capabilities of their KABAFA. When a person is taught about the beauty of blackness and the depth of darkness as positive ways to conceptualize reality, there can be a magnificent change in consciousness. As the consciousness of the person changes, you then impact the spirit. On this level of change, both a nation and a people can change if the therapy is focused on what was in the beginning. Did your day not start in darkness? As it was in the beginning, so shall it be in the end.

CHAPTER 7

KABAFA – The Flesh of Ra

Fig. 15 For RA, it is the ancient symbol for the sun.

The Godfather of Soul (James Brown) sang it and the hip hop artists (Erik B. and Rakeem) sprang it:

I know you got soul.
If you didn't you wouldn't be in here
I know you got soul
If you didn't you wouldn't shake that thing.

Everyone on the planet has the power to display the functioning of KABAFA. From the Eskimo in the Arctic to the indigenous people in the archipelago of Chile to the African, Asian and European in any region of mother earth, the sun enriched physiological functioning of KABAFA is affecting reality.

Having KABAFA is like having spiritual swagger. KABAFA is the rawness of reality. It is the human being uncut, with no cosmetics, special haircut, or clothing to demarcate reality. Whether melanin-dominant or melanin-recessive, the expression of KABAFA keeps the individual receptive to the spiritual realms of existence.

Scientifically, the terms eumelanin, pheomelanin and neuromelanin will still be in use because the chemical content and structure of each type of pigment is slightly different. The complex pigmentary processes cannot be limited to just a word as some emphasize that melanin should be correctly called carbon (Kamene, 2024). I have already written on the science of melanin, so the scientific principles stand as a legitimate area of study. The polemical debate will be on the discussion of how to interchange melanin, carbon and KABAFA. As long as we establish that pigmentation is a scientific advancement in nature, it does not matter what you call it. In our society, blackness, darkness and any entity which is heavily pigmented is thought to be negative. This slanted view is basically propaganda that is globally espoused by those who are melanin-recessive. It is emphasized here that being melanin-recessive makes you less likely to fully display KABAFA on a psychospiritual level.

Fig. 16 TiMo in front of the Great Pyramid showing the KA symbol.

In the human body, we could say the word KABAFA as a replacement for melanin in the skin, hair, eyes, internal organs and the brain. In contrast, we do not necessarily need to say that KABAFA is the replacement name for every pigmented cellular group that is found, but KABAFA is particularly the expression of the collective nature of all pigmented systems that make you a spiritual being having a human experience. From a cellular level, we have the term melanogenesis to describe the embryological development of pigmentary systems in the body. Therefore, we might not need to get into" kabafagenesis" as an interchangeable term. Pigments are in every aspect of life, so pigments and the use of melanin will likely not be eliminated from our lexicon. The way history works, it could take another 100 or 1,000 years before the intellectual world can reconceptualize what is proposed in this book. For now, we must live for the current day. On a sociopolitical level, the language of using KABAFA can change policing, legal decisions, sports commentary, social relations, interpersonal dynamics, politics and economic affairs. It is critical that we impact future generations for success in fighting the spiritual warfare that is being waged. The power of your KABAFA will prove that the spirit of a people can conquer all flesh.

The Flesh of Ra

The Flesh of Ra has been revealed in KABAFA. Melanin-recessive people do not fear the actual person who is pigmented, but the fear is in the inevitable genetic annihilation of interacting with the masses of people who display KABAFA. It may not seem to be on a conscious level, but this fear of KABAFA has organized melanin-recessive people to globally despise the Flesh of Ra. Other people's fears can be dangerous to the way people interact, but people who have the Flesh of Ra must use the spiritual energy of KABAFA to fight this negative onslaught. KABAFA has the power to change the world on many levels, so let us be spiritual warriors to heal humanity.

Imbalanced power relationships have developed because of the incessant claim that some people (white) are superior to others (nonwhite) in the socio-political landscape.

The inferiority projected onto people with KABAFA was never that they were physically inferior, but there had to be a centuries campaign to project melanin-dominant people as mentally inferior to make melanin-dominant people feel subordinate. Once the mind is taken, the body is controlled. As we rise above the Flesh of Ra and experience the natural talents of our essence, we can create a new humanity. We have done it before, and we can do it again. We are not stuck in this flesh. Like the rays of Ra, we can expand and reach parts of the universe that have not been touched yet. So be a human ray of sunshine and be the light for the world.

The importance of this concluding chapter is to reflect on the human condition and how humanity can only change when people understand the spiritual nature of all humans. KABAFA is everywhere, and it is the definition of a peaceful reality with nature. Just because you may possess KABAFA does not make you superhuman or overly righteous because evil lives everywhere, like KABAFA is present everywhere. It is critical in this conclusion to focus on why we need a change in consciousness to truly understand these spiritual concepts.

Energy and Inflammation Balance

The electricity in our consciousness is the spark of life. The ideas that emanate from our mind are what create the entire reality of what we see. Therefore, we cannot limit melanin as the sole entity to define what collectively binds all elements in nature. For KABAFA, we are discussing human nature. Since we have melanin, blackness, carbon or pigmentation in every aspect of nature in the cosmos, we need to look at this phenomenon in another perspective when we delve into human nature.

Darkness really means life, for out of the darkness we find light. Black and dark usually relate to something bad from a Eurocentric outlook, but this negative thinking is not what is being referred to. We must create a new way of viewing this Greek Eurocentric word called melanin. As presented here, all old concepts related to melanin will now be called KABAFA instead of melanin when describing human nature.

In its most elegant simplicity, life is about energy (E) and inflammation (I). Energy is everywhere, and all life forms that depend on electromagnetism are able to channel energy for specific needs. When a life form becomes injured or hurt, inflammation occurs to mend the life force. The balance between energy and inflammation is nature. In other words, when nature is in balance, life exists perfectly. In the grand scheme of things, nothing is perfect so life waxes and wanes between energy and inflammation.

The E-I concept has been established for a millennium. Yes, for thousands of years of human existence, the manifestation of life has been to maintain this balance. If we juxtapose life today in the 21st century, artificial intelligence or AI is becoming the dominant concept. AI, however, is dependent upon the programmer. The efficiency of AI technology can alter the balance we seek in nature, and the alteration has both pros and cons.

The human brain can make mistakes, but the computer is modeled for perfection and accuracy. When data is analyzed and there appears to be inconsistencies, more than likely, there is a data entry error. With the world now dependent on electricity for data to be processed, human input is becoming less needed. People are becoming obsolete, and jobs are diminishing. AI, in the form of robots, is running the show and the quality of human interactions has dissipated.

Harnessing your KABAFA is an approach to counteract the onslaught of AI on the psyche and return to E-I for balance. KABAFA is the original intelligence that has no artificial label. Your KABAFA organizes cells, promotes growth, mends tissue, protects against radiation and toxic substances, extends life, converts energy, connects different realms of existence, and it exists in inanimate elements. KABAFA can exist without humans, but AI would not exist without humans.

We can see the importance of chlorophyll as a pigment for life and plants, but seldom do we recognize the similar role of KABAFA in humans. Every principle you think about chlorophyll in plants applies to KABAFA in humans. The suggestion here is that it does even more because plants do not display the same level of creative intelligence as humans. The KABAFA in humans provided the basis for advanced thought in the human experience. KABAFA has elevated humans beyond the animal-like presence that pervades the planet.

Historically, someone had to step away from the animal-like state and become the originators and creators of civilization. The original KABAFA-dominated appearance of the original humans on this planet were dark in color and full of KABAFA. Those people still exist, although admixture of genes have dispersed throughout earth. The originators are now despised, but through proper education and not indoctrination, we can restore the history of those who have been perpetually exploited or denigrated.

KABAFA is the physical presence of spiritual elements in the body. The body is nothing but a vehicle for the soul to travel. As the soul travels, the mind or nous overwhelmingly influences the direction of the soul. Therefore, as we break down KABAFA and what we have commonly called melanin, we are speaking of the science terms that allow KABAFA to enhance the sensorimotor

experience of life. The energy that pervades the universe can be tapped into with the combination of the body, the breath, and the power of natural pigmentation inside and outside of the body. Tapping into your KABAFA will ultimately energize your life.

The colonizer has exploited earth's resources by tapping into the core of the planet. Oil drilling, fracking for methane gas, and digging for precious metals are examples of exploitative industries mastered by the colonizer. The same way soulless people can tap into the earth to gain energy from these natural resources, these same soulless people can dig into the human being and extract their mind, body, and spirit. By explaining a greater role of pigmentation in life, we can redefine the limited use of melanin to explain cultural variations in studying the mind, body, and spirit. In other words, you may not view the world like another person because you may or may not have a connection to the cosmos. Whether KABAFA-dominant or KABAFA-recessive, we have people of all hues who will throw trash out their car window and pollute their immediate surroundings. This is the ultimate display of a person who has lost their connection with nature. Their KABAFA is contaminated with the wetiko virus (Forbes, 1992).

In conclusion, in ancient Kemet, the sun was known as RA. RA was worshiped during certain periods in history, and it remains the solar generator in our little galaxy. As a burning star, RA emits RA-diation that can be felt in many types of frequencies throughout the electromagnetic spectrum. Our naked bodies developed to absorb this RA-diation and our skin has the unique flesh that RA saturates to impact our internal organs and overall consciousness. As we have broken down the KA, BA and FA of this collective name, we can emphasize that the pigmentation in and on our physical bodies is the flesh of RA to enhance our KABAFA. Know thyself and allow your KABAFA to fully manifest by embracing the RA-diation to experience the fullness of life. KABAFA is melanin poppin'.

Epilogue

The KABAFA Story

One day an eagle egg fell out of the sky from a hole in the nest, and it dropped into the water and drifted into a ravine. Then along came a fox looking for food, and it found this lone egg that was bigger than other eggs the fox had seen before. The fox was collecting stolen eggs from the local farmer, so it added this egg to the pile of chicken eggs that were stored in a secret location.

One day, the irritated farmer snuck up on the sly fox and chased it away from the area in which the fox had been stashing the eggs. The farmer carefully returned all the eggs back to the farm and then placed all the eggs back in the chicken coop for the mother hens to provide warmth.

After a few days, all the eggs broke, and the birds hatched. They steadily grew over the weeks and displayed bird behavior, but this one bird seemed different. It was bigger, stronger, and it had more KABAFA (i.e., the feathers were colored different to absorb the heat, the bones were stronger for greater movement, it mentally developed quicker, it had a keen awareness of nature, and it was more agile).

Even though all the chickens were complacent with just walking around in circles, this different bird had an innate desire to look up, there was greater extrasensory perception, and there was more of a connection to nature beyond the chicken coop. For this strange colored bird in which the white chickens called him Uncle Tom, there was an internal desire to swim, dive, hunt, and soar. However, this colorfully strange bird did not know how to express the inner essence of what he could truly be. Until one day, Uncle Tom looked up in the sky and saw a similar looking bird doing what he never thought chickens could do. As this glorious looking bird flew high in the sky, it looked down and asked, "What are you doing stuck with them chickens, those foxes are going to get you?"

This majestic bird swooped down and perched on a pole next to the chicken coop. In amazement, this glamorous looking bird (H.T. for Harriet Tubman) said to the confused bird (H.T. for Herschel Thomas), "What are you doing down here with these white chickens?" Herschel Thomas replied, "because I am a chicken, I just lack the white color." Harriet Tubman responded with an irritated look, "Fool, look at your reflection in the farmer's mirror, you ain't no chicken. You are a majestic eagle, now get up here and fly."

HT chicken – "I can't fly, I can't soar, I can't do what you do in the air and beyond."
HT eagle – "Who told you that?"
HT chicken – "My fellow chickens and that boss man farmer on the plantation."
HT eagle – "You stay stuck with them chickens and that overseer, and you will get eaten by them foxes. Spread your wings, extend your legs, open your inner vision and feel the power of the creator in you. Move to the rhythm of life and stop walking around in circles with them chickens."
HT chicken – "You mean I can fly?"
HT eagle – "You can do more than that. You can glide in the sky and live in harmony with the cosmos. Also, you can spot them foxes from way up in the sky."

After the belief was generated and the self-knowledge was fulfilled, the world was about to change for HT chicken. He was encouraged to leave them chickens and create his own world after hearing the powerful words of HT eagle. HT eagle boldly said, "Fly with me and honor your genetic heritage. KABAFA is in you."

What this story is trying to reveal is the miseducation that occurs when people do not know their heritage. In ancient African history, we can look in amazement at the physical structure of massive monuments that were built without modern tractors and cranes. We can try to visualize the special nature of an ancient people who used a different technology than today to build these massive structures. For example, the construction of pyramids and temples in Africa is incomprehensible to the modern mind. Our greatness has been suppressed, but the evidence of what humans have done is still standing. As we look at the physical prowess of modern-day athletes and the unbelievable psychomotor skills displayed by humans with higher levels of KABAFA, we begin to understand the inner power we are describing. Just having pigment in your skin is not what drives your behavior. It is the overall presence of KABAFA that is encapsulated in your genes, and the more biophysiological darkness (capacity to produce melanin) in the genetic makeup can highlight the full expressiveness of KABAFA.

References

Akbar, N. (1994). Light from Ancient Africa. Mind Productions & Associates: Tallahassee, FL.

Baines, J. (2008). The Secret Science. John Baines Institute, Inc: FDR Station, NY. (First published 1980).

Bauval, R. and Brophy, T. (2013). Imhotep The African: Architect of the Cosmos. Disinformation Books: San Francisco, CA.

Ben-Jochannan, Y.A.A. (2005). Understanding the African Philosophical Concept Behind the Diagram of the Law of Opposites. Black Classic Press: Baltimore, MD. (First published by Alkebu-lan Books and Associates 1975)

Carson, B. (2019). Compendium of the Emerald Tablets: A Beginner's Guide. 4biddenknowledgeinc: Coral Springs, FL.

Clark (X), C., McGee, D.P., Nobles, W. and Weems, L. (1975). Voodoo or I.Q.: An Introduction to African Psychology. Journal of Black Psychology, Vol. 1, No. 2, pp- 9-19.

Forbes, J. (1992). Columbus and Other Cannibals. Autonomedia: NY.

Gates, H.L. (2024). The Black Box: Writing the Race. Penguin Press, NY.

James, G.G.M. (2005). Stolen Legacy: Greek Philosophy is Stolen Egyptian Philosophy. Khalifah's Booksellers and Associates: Drewryville, VA. (Originally published 1954).

Kamane (2024). Carbon Before Melanin: Creators on Earth. Panther Prince Productions.

Malz, B. (1977). My Glimpse of Eternity. Chosen Books: Waco, TX.

Menzise, J. (2015). Symbolically Speaking, Vol. 1: African Lodge #1-The Context. Mind on the Matter Publishing: College Park, MD.

Moore, T.O. (2020). Pigment Power: Topics on Melanin in Science and Health. Zamani Press: Redan, GA.

Moore, T.O. (2022). Fire Atoms Ignite Inner Vision. In Eds. E.B. Bynum, A.C. Brown, R.D. King and T.O. Moore, Why Darkness Matters: The Power of Melanin in the Brain (3rd Ed), pp. 161-172. K&R Printing and Publishing, Ambler, PA.

Nelson, T. (2003). KaAbBa: Building the Lighted Temple (2nd Ed). The Academy of Kemetic Education & Wellness, Inc: Mattapan, MA.

Nelson, T. (2014). MerKaBa: The Great Pyramid is the Tree of Life: KaAbBa (3rd Ed). The Academy of Kemetic Education & Wellness, Inc: Mattapan, MA.

Papas, A.M. (Ed.) (1999). Antioxidant Status, Diet, Nutrition and Health. CRC Press: Boca Raton, FL.

Seiss, J. A. (1973). The Great Pyramid: A Miracle in Stone. Rudolf Steiner Publications: Blauvelt, NY. (Originally published 1877).

Tolokun, O.Y.A. (2020). Afrikana Spiritual Coaching: A Cultural to Actualization. Sakhu School Publications: Atlanta, GA.

Van Sertima, I. (1989). Egypt Revisited: The Journal of African Civilizations, Vol. 10 Summer 1989. Transaction Publishers. New Brunswick, NJ.

Glossary

A

Afterlife
The afterlife means what some people believe happens to the soul after the body dies.

Amenta
In Kemetic teaching, Amenta is the hidden 'underworld'—a place linked to the unconscious mind and deep inner feelings.

Ancestors
Ancestors are family members from the past. In the book, people may honor them and ask for guidance through prayer or rituals.

Ancient Kemet
Ancient Kemet is the ancient name for Egypt. It means "Black Land," referring to the rich dark soil along the Nile River and it's melanin-rich people. In the book, Kemet is used to talk about African history, spiritual science, and ideas about the soul

Ankh
The ankh is an ancient Kemetic symbol that stands for life, breath, and life continuing beyond one lifetime.

Astronomy
Astronomy is the study of space—things like the sun, stars, planets, and the universe. In the book, it connects to how the sun and space energy relate to life.

Ayahuasca
Ayahuasca is a traditional Amazonian drink made from plants that can strongly change thoughts, feelings, and visions. The book discusses it as a type of plant medicine used for healing, initiation, and exploring consciousness.

B

BA
BA is one part of the soul in KABAFA. It relates to breath, oxygen, and the life force that supports the mind and body.

Bantu-Kongo Cosmology
African worldview describing humans as radiant energy beings.

Black Dot
Symbol representing inner consciousness and spiritual potential.

Brain
The brain is the organ that controls thinking, memory, and consciousness. The book links certain brain areas (like the brain stem) to deep inner awareness and pigments like neuromelanin.

Breath
Breath is the air you move in and out of your lungs. The book explains how breathing affects focus, feelings, and energy.

C

Carbon
Carbon is a basic building block of life found in many living things. The book connects carbon and pigmentation to how the body handles energy and protects cells.

Cells
Cells are the tiny living units that make up your body. The book explains that cells work together, and pigments can help rotect cells and support how the body functions.

Chakra
Energy centers influencing health, emotion, and consciousness.

Consciousness
Awareness shaped by mind, body, breath, and spirit interactions.

Cosmology
Cultural understanding of universe, energy, and spiritual balance as taught in African traditions.

Creation
Creation means how something begins or is made. In the book, it can mean the start of life and the universe, and how mind, energy, and nature work together.

D

Dark Energy
Dark energy is a name scientists use for an unseen force in space that seems to push the universe to expand. The book compares nseen forces in space to hiddenspiritual realms.

Dark Matter
Dark matter is matter that scientists think exists in space because of its gravity, even though we cannot see it directly. The book connects this idea to unseen parts of reality.

Dark Universe
Scientific concept of dark matter and energy linked to African cosmology of unseen realms.

DMT
DMT is a chemical sometimes called the 'spirit molecule.' The book connects it to dreams and strong vision experiences

Dreams
Dreams are stories and images your mind experiences while you sleep. The book uses dreams to discuss inner vision and how the

mind can experience things beyond ordinary time and place.

Dya
Bambara concept of life energy unifying the natural world.

E

Electromagnetic Spectrum
The electromagnetic spectrum is the full range of energy waves, from very strong cosmic rays to radio waves.

Electromagnetism
Energy framework fundamental to life, consciousness, and KABAFA functioning.

Energy
Energy is what makes things move and work. The book also talks about energy as a life force in people and nature.

Enzymes
Enzymes are proteins in your body that help chemical reactions happen faster, like breaking down food or protecting cells. The

book mentions enzymes that help defend the body from damage.

Evolution
Evolution is how living things change over long periods of time. In the book, it is mentioned as part of how life develops and adapts.

F

FA
FA means 'fire atoms'—tiny spiritual-energy particles described as part of creation and consciousness.

Fire Atoms
Ancient concept of subtle spiritual particles forming mind and soul; reinterpreted as FA.

Flesh of RA
"Flesh of RA" is a way to describe how the sun (RA) powers life through light and radiation. In the book, it connects the sun's energy to pigmentation and spiritual strength.

Free Radicals
Free radicals are unstable molecules that can damage cells. The book says antioxidants help protect the body.

Frequency
Frequency means how fast something repeats, like a wave or vibration. The book uses frequency to explain how energy and nature can affect the body and consciousness.

G

Glutathione
Glutathione is a powerful antioxidant in the body. It helps clean out toxins and protect cells.

H

Healing
Healing means getting better after injury or sickness—physically, emotionally, or spiritually. The book talks about healing through balance, breath, and sometimes plant medicines.

Heart
The heart is the organ that pumps blood through your body. The book mentions the

heart as part of life energy and how the body's systems work together.

Human Body
The human body is your physical form—organs, cells, and systems that keep you alive. The book explains the body (KA) as part of the full KABAFA experience.

I

Inflammation
Healing response that must balance with energy for well-being.

Initiation
Initiation is a learning process where someone is trained in spiritual knowledge, discipline, and self-control.

Inner Vision
Inner vision means "seeing" with your mind—intuition, deep awareness, or spiritual insight. The book connects inner vision to the third eye and expanded consciousness.

K

KA
First soul division representing the physical body as vessel for spiritual energy and electromagnetic absorption.

KaAbBa
Term explored by Nelson describing body–spirit energy vehicle similar to KABAFA.

Kalunga
Bantu concept describing universal life force circulating through all existence.

Karnak
Ancient Egyptian temple where initiates studied spiritual sciences.

Kra
Akan concept for personal life force and destiny.

L

Life Force
Life force is the energy that makes living things alive and active. The book describes life force as connected to breath, spirit, and the body's ability to handle energy.

Light
Light is energy that travels in waves and lets us see. The book connects light from the sun to pigmentation, the electromagnetic spectrum, and how the body absorbs energy.

M

MAAT
MAAT means truth and balance. It is about living in a fair, honest way that keeps life in order.

Meditation
Meditation is a practice of calming the mind and focusing attention, often using the breath. The book connects meditation to balance, consciousness, and spiritual growth.

Melanin
Melanin is the pigment that gives skin, hair, and eyes color. In KABAFA, it is also described as helping the body handle light and energy.

Melanin-Dominant
Term describing bodies with higher pigmentation and heightened sensory–spiritual integration.

Melanin-Recessive
Lower pigmentation affecting sensitivity to natural frequencies; contrasted with KABAFA expression.

Memory
Memory is your ability to store and remember information. The book links memory to brain function and pigments like neuromelanin.

MerKaBa
MerKaBa is described as an energy 'light body' that supports spiritual growth and inner travel.

Mind
The mind is your thinking and awareness—your thoughts, emotions, and focus. The book explains that the mind is central to wellness and connects to spirit and the body.

N

NDE

NDE means near-death experience—when someone feels they briefly left the body during a life-threatening moment.

Neuromelanin

Neuromelanin is a dark pigment inside parts of the brain. The book links it to memory, intuition, and deep inner processing.

Nyama

Dogon concept of internal spiritual energy flowing through humans and nature.

O

Origin

Origin means where something starts or comes from. The book uses origin to discuss where ideas, life, and spiritual traditions come from.

Oxygen

Oxygen is a gas you breathe in that your body needs for energy. The book explains that oxygen must be balanced because too much can harm cells, and the body has ways to protect itself.

P

Perception
Perception is how you understand what you see, hear, and feel. The book explains that perception can change based on consciousness, inner vision, and spiritual awareness.

Pineal Gland
The pineal gland is a small part of the brain. It helps with sleep chemicals and is linked in the book to intuition.

Plant Medicine
Natural substances that enhance spiritual insight, emotional healing, and mind–body awareness.

Prayer
Prayer is speaking to or connecting with a higher power through words or thoughts. In the book, prayer can be part of rituals and spiritual focus.

Pyramid
Ancient energy structures used for initiation, burial, and cosmology.

R

RA

RA is the sun power in Kemetic belief. It represents life-giving light and energy from the sun.

Radiation

Sunlight and cosmic energy absorbed by skin and internal pigments.

Ritual

A ritual is a repeated action done for meaning, prayer, or tradition. The book describes rituals as ways to focus the mind and spirit.

S

Science

Science is a way of learning by observing, testing, and explaining how things work. The book blends science ideas with spiritual concepts to explain KABAFA.

Soul

Spiritual essence of humans expressed through KA, BA, and FA interactions.

Soul Therapy
Practice guiding individuals toward inner healing, balance, and emotional clarity.

Spirit
Invisible life force permeating body, nature, and universe; eternal beyond physical death.

Spiritual Energy
Spiritual energy refers to the life force or inner power that connects the body, mind, and spirit. In the book, it describes the unseen energy that supports consciousness, healing, and connection to nature.

Spirituality
Human connection to energy, consciousness, and universal forces beyond physical perception.

Sun
The sun is the star that gives Earth light and heat. In the book, the sun (RA) is described as a main source of energy that supports life and affects pigmentation.

Superoxide Dismutase
Superoxide dismutase is an enzyme that helps protect cells from damage caused by reactive oxygen.

T

Third Eye
The ‘third eye’ is a way to describe inner vision—feeling and understanding things beyond normal sight.

Thoth
Thoth is a Kemetic figure linked to wisdom, writing, and knowledge.

Transformation
Transformation means a big change. The book uses it for personal growth—changing how you think, act, and understand yourself.

Trauma
Trauma is a hurtful experience that can leave a strong emotional mark. The book discusses healing and finding balance after trauma.

U

Underworld
The underworld is a hidden realm in many traditions. In the book it relates to unseen life, spirit, and deep inner mind.

Universal Consciousness
Universal consciousness is the idea that all life is connected through a shared awareness or spiritual reality. The book links this to energy, spirit, and the unseen world.

Universe
The universe is everything that exists—space, stars, planets, and energy. The book connects the universe to cosmology and to unseen forces like dark matter and dark energy.

V

Vision
Vision is the ability to see, but the book also uses it to mean understanding and inner sight. It connects vision to awareness, the third eye, and spiritual insight.

W

Water

Water is a basic need for life. The book mentions water as part of nature's balance and as a symbol connected to life and spiritual ideas (like NUN).

Wisdom

Wisdom is knowing how to make good choices using knowledge and experience. The book connects wisdom to learning, balance (MAAT), and spiritual understanding.

Index

A

B

C

D

E

F

G

H

I

K

L

M

N

O

P

R

S

T

U

V

W

www.ingramcontent.com/pod-product-compliance
Ingram Content Group UK Ltd.
Pitfield, Milton Keynes, MK11 3LW, UK
UKHW022029041026
15591UKWH00012B/583